ANIMAL EXPERIMENTS

Anja Konig grew up in the German language and now writes in English. Her pamphlet *Advice for an Only Child* was shortlisted for the 2015 Michael Marks prize.

Animal Experiments

Published by Bad Betty Press in 2020
www.badbettypress.com

Cover design by Amy Acre

Printed and bound in the United Kingdom

A CIP record of this book is available from the British Library.

ISBN: 978-1-913268-05-3

Supported using public funding by

ARTS COUNCIL ENGLAND

LOTTERY FUNDED

ANIMAL
EXPERIMENTS
ANJA KONIG

PRESS

Chuang Tzu and Hui Tzu were walking beside the weir on the River Hao, when Chuang Tzu said "Do you see how the fish are coming to the surface and swimming around as they please? That's what fish really enjoy!"

"You're not a fish," replied Hui Tzu," how do you know what fish enjoy?"

Chuang Tzu said, "You're not me, so how do you know I don't know what fish enjoy?"

—Master Chuang, The Great Happiness, 17

Animal Experiments

Contents

iii. the things that make me happy make me sad

iv. does the butterfly remember what the caterpillar knows?

Cat and Lizard

I wish
I had not
seen that

i.

this long moment, life, suspended between not yet and no longer

Solstice

I cross
the bridge
in the early dusk.

The things
that make me happy
make me sad.

You won't see
Midtown glow
through the grid,

the Navy Yard piled
high with grit, the sugar
factory half gone.

You will never eat
eight dumplings
at C&C Prosperity

on Clinton Street.
These are dark days.
I wait.

Triple Negative

We always met at Cafe Blunt—
latte, eggs, Moroccan tiles.
Your hair was growing back
more wiry and wild.
      You said that it had spread—
brain, liver, bones,
a butcher's plate.
You looked afraid. We talked
      of other things,
that we should get out more,
enroll in match dot com, how hard it was
to make new friends.

Fall Time

The last cricket
drowns
in a bucket
of brackish water.

Parasitic hairworm
proteins invade
the geotactical center
of the host brain.

Springloaded jewelweed
fires seeds
into the rain.
It's time.

The stove runs cold.
I pour out the oil,
bury the gold,
pitch the tarpaulin.

I eat bracken
for vitamins.
I sleep
on my knife.

Pork Parable

After Master Chuang

Science tells us that she [a sow] doesn't even seem to know
that she can't turn... She wants to eat and feel safe, and she
can do that very well in [a gestation crate].

—Paul Sundberg, National Pork Board

I'm not Paul Sundberg.
I can't know if he knows
what a sow really wants.
For all I know, a sow
likes life in a crate

too small to lie down,
desires nothing more
than to give birth, be fed
and shit through slits
in the concrete floor.

Paul Sundberg knows
a sow's not meant to think.
She is content
to be contained
inside her private stall

where she feels safe.
She isn't interested
in entertainment.
Turning around
is overrated.

Grateful, safe
until the steel bolt
to her head. It is humane,
she doesn't
feel a thing.

Tenderness

Pork.org found tenderness
is most important.
Five genes can lead
to pigs producing
tough pork! Research
is needed to determine
the association of these genes
with other traits, like growth
and lean meat yield.
Tenderness is hard to get.

The Storks

are back,
ten of them
piercing frogs
in the mud.

As kids
we were taught
they were
extinct.

Things
may be more
resilient
than we think.

Therapy

We have a ritual:
when I call she says
why don't you call.

Next we talk about
my poor performance
as daughter.

You should be grateful,
she says, *other parents*
kill their children.

I know she is right.
In the worst childhood
competition my mother wins.

She barely survived
the village
by the Czech border.

They had blankets
instead of doors.
One in four went mad.

The dog was scalded
in boiling water
and had to be shot.

Uncle Rudi smashed
the drawers shut
on her baby fingers.

The priest called her bastard
in front of everyone
on Sunday.

I ask what her therapist thinks
about it all. *That's none
of his business*, she says.

Animal Experiments

She was relieved it was a girl,
because genetic aberrations
occur more commonly in boys.
She had grown up beaten by nuns
and wanted it better for me:
food on the table, a roof.
Nobody mentioned love.

Finally Peace

your mother says
as she drinks the Pentobarbital.
I go to see her at the morgue
in her white coffin, stern and small,
a box of tissues on the single chair.
I guess that is a kind of peace.
The plastic brace around her neck
cuts into her chest.
When I look closely, I see
the black tip of her finger.
She is wearing red sandals,
her tiny toes are made of wax.
I take a photo.
That's what you would do,
if you were alive.
Fuck peace.
I want tears. I want life.
Monday a beautiful ghost speaks to me
with a silver voice
and kisses me deeply.
There is greed in his eyes
and panic. Wednesday he dumps me.
I want more
of this foolishness, false hope,
the sleepless hammering of the brain.
Every day this sweltering summer
I swim against the Limmat,
every day I think of you
out in the thunder.

Body of the Beloved

I think you would like
being described
in the police report
as *slim male corpse*.
There is no crepitation.
Lividity is appropriate
to your position.
Your light brown hair
is mentioned, your eyes,
slightly open. Rigor
of the jaw locks your lips
against the finger
in the latex glove.

Not the Last Chapter

I'm up all night,
trapped
like a bottled bug.
London is hardest.
The broken escalators
we used to mock
can bring it on.
The 24 is stalled
opposite the curry house.
The driver smokes one outside,
passengers look on.
Attachment is suffering.
At Southbank
pedestals are crammed
with trash. Planks jut
into the river where I took
the call. The bridge,
the brink, the wall.
Cats don't get insomnia.
We can learn from that.

I Nearly Took the 24 Today

Remember, once it just stopped,
the driver got out and walked off.

Remember, on the left side of Lupus
a pedestrian might get hit by an empty beer can.

Remember the mould, the leak, the water bills,
the boiler. I thought we were unhappy then.

Remember gooseberry pickles, like eyeballs
in a jar, you said they help a person relax.

Remember when it snowed for three days and
I did not have to go to the airport.

ii.

if you don't belong to where you're from, you can make anywhere home

Vogelfrei

Ginger: We'll either die free chickens or we die trying!
Babs: Are those the only choices?

—*Chicken Run*

Once you've flown the coop,
why stop, give up the open road,
return to ground?

Once you've taken orders
only from yourself,
why settle down?

Once you've seceded
why keep the queen?
Go all the way!

Once you've succeeded,
drilled your own oil,
minted new coin, why not

dissolve your lonely country,
disperse the tribe,
dismiss your outer islands,

count your only vote,
elect yourself and sail
a single acorn into space!

Undoing

I return
to all the places.

Everything I have done
I have also undone,

marriage,
citizenships.

Trying on clothes
nothing fits.

If you don't belong
to where you're from,

you can make anywhere
home.

Little *viola cenisia*
root shallow and fast

on shifting scree.
Hang on.

Metamorphoses

Ants in their city of sisters
sing molecule lullabies

    *

my animal heart
my nettle nerves

    *

five years underground, four weeks on the wing
stag beetle, sing!

    *

blind architects, boys and girls in desert towers
termites eat the world, peaceful little ones

    *

does the butterfly remember
what the caterpillar knows?

Animal Experiments

Now that I've seen the first one,
I see them everywhere this spring:
pushing prams, holding hands with toddlers
on the tram, babies in the Baby Björn,
shooting ball with sons and daughters.
Once you're used to it, it seems quite natural.

I have vague memories of mine
in the chair, watching news on Bayern Drei.
Once, when I was eight, we went for a walk
in the leafless park, just the two of us, embarrassed
by the lack of conversation. My mother says
he's not the one who taught you how to speak.

Old-style, mostly inert. I remember
my tantrum at three, when he wouldn't hold my hand.
Nothing too bad. He once pushed the pram
into the moor and walked away. I forget to call,
sometimes for months. The young fathers
are all over the city. I can't stop looking.

Saturday 6am

Having just
come in
from the already
bright terrace
together
with traces
of meadowy air
and hyacinth
sitting down
on the chair
with the one
plate and the one
cup in front of me
I see the day
is mine
entirely.

Tragedy

With affection
you can train a beagle
to offer his paw
for the injection.

Captive

A bunch of monkeys tumble
out of crates
after the sedated journey
from the jungle.
Right away they ask
The Question:
who's on top?

I read it's true
for any social species,
chickens or beagles,
or people.

In any colony of captives
there's always an alpha
in the cage,
then all the others
down the power ladder.
To the guy with the syringe
it doesn't matter.

Internship with Kim Jong-Il

For fifteen months I took the minutes
in his committee. His direct reports
came to the conference room
        in tears. Some were shaking,
rattled by the travel and the waiting.
When opening the door, I did it gently,
made my smile kind. Some
        knelt and begged, some lost
their heads. Men, who in turn had other men
who knelt and begged in front of them.

Animal Rights

At dinner the other night in one of those places
in Westminster where I would never go if it weren't for business
we were twelve guys after work drinking beer,

and me. A few beers in,
one of them, the pretty one,
in his heart a nice man, but confused

by many new things in his life, his wife for example
asking him to put the kids to bed.
A man with an MBA,

who nonetheless didn't know what, or who
King Lear was, someone who likes the sense of community
at his Firm, the golfing and the moneymaking,

not a hard man, an ordinary fellow, enjoying his ordinary privilege,
his whites from South Africa and his reds from Sonoma,
this man said, why bother with the PhD, why did you not

for example become a stripper, and he laughed. Clearly
he was only teasing, he said, and by the way,
how could a dog or any animal have rights, same

as a human, but who knows nowadays:
even women have rights. He was just teasing.
I smiled, so he would know

that I, too, was only teasing, and offered
to castrate him with my fork; nobody shall call me
humourless.

Lucky Logan

Logan doesn't know he's lucky
good boy
how big will he grow?
three days ago at the pound
his time was running out
when this big bald bearded guy
I met on Pinsker Street
adopted him
from death row
to dog about town
Logan doesn't know
he's lucky

Cats and Dogs of No Esteem

... seeing these effects will be
both noisome and infectious

—*Cymbeline*

Nude mice carry our tumors under their
skin. Beagles are doomed by their human
hearts. Cynomolgus monkeys can be killed
only carefully. The predominant emotion of a
mouse is fear. Higher animals such as humans
also know curiosity.

Charismatic Megafauna

for Gus, the bipolar bear of the Central Park Zoo

Gus
was
just
like us.

He had
a pad
as small
as some
Manhattan
flats.

Gus
was
sad.
He saw
someone
for that.

Gus
had swum
alone
for far
too long.

Gus
grew
old
but
died
too young.

Gus
was
just
like us.

We are the Bees of the Invisible

says Rilke, of poets (I think). I thought I could never
be a bee: all that peer pressure, the hum
and hustle of the hive. Who can relax
when the next bee is doing her urgent dance?

But the bees of the invisible live wild,
solitary lives—in bee hotels in botanical gardens,
in hollow reeds, in holes in dry bricks,
a kind of Manhattan at the edge of a wood.

Each singular bee in her cell makes
honey, so transparent that it looks like nothing.
But a hungry tongue can just detect its mix
of tastes, its texture, tough and sweet, like love.

iii.

the things that make me happy make me sad

October

The world being
praiseworthy
I went
and in the early morning
bought a pen.

Happiness in Dark December

The scent of viburnum carries me over,
its mad winter blossoms by the tram stop.
Everything pleases me today:
the little city by the lake,
the café with rose-covered wallpaper,
eggs, creamy OJ, Celine taking a selfie.
I climb the hill I have sometimes been bored with.

My work pleases me, to be a citizen of any place,
this long moment, life,
suspended between not yet and no longer,
the body without pain except
a twinge in my left hip as I climb higher
towards the setting sun, out over the fog
under the mountains shocking with snow.

The View from the Top

Junko Tabei has died in Japan.
When she stood on the summit
of Everest she was glad
she didn't have to climb any higher.

I think of Sunaiyanaa
whose father didn't want her to drive.
She found a cabbie in Bangkok, paid him by the hour,
passed the test, moved north.

I think of Rosalind who loved to climb the Alps,
who was meticulous, alone,
who photographed life itself.
They wouldn't call her Dr. Franklin.

I think of the girl whose name I don't know
walking to school afraid of acid.
I think of the mutilated girl, of her mother, too,
who is holding her down for the knife.

I think of the women who climb through ice,
some shattered by falling seracs,
some slipping silently into crevasses.
Some make it across the ladders.

Women's Pond

moss brown water
plunge in
algeous hair
breathe drift
shores teem
with bones breasts
steaming in the sun
a pitbull's lolling tongue
a squirming baby
under a chiffon shroud
35 degrees on the Heath
melting orange slush
from the boxed and furious
ice cream truck
women women women
crowd the water's edge
six deep
slide in
one by one by one
none scream

Controlled Substance

it is the woman's part
    —*Cymbeline*

Diaphanous girls
        like henbane saplings
are bent to the lattice
        for the raptors' harvest.

Residual elders
        rule errant bodies,
stoppered
        lest precious issue
                    escape.

The Fathers' furthest root
        runs into the under rock.
At one part per million
        women's vermilion
                can contaminate
the subterranean
Antarctic lake
        and colour it in
                    blood.

Your Heart Does Not Want to Sleep

In the country under water
the clouds are listening.

Translucent immigrants,
with only traces of orange,
spend nights watching for leaks.

Women wear chalk on their lips,
men walk on illuminated feet.

Citizens patrol the dams in shifts.
The moonsheep breathe, cluster.

An enticing type of parasite
feeds on their wool, turning it slowly

the colour of nothing. Some say
wool is more valuable like that,
although the sheep are infected

by a languor so deep
their meat cannot be digested.

Extinction Rebellion

for Jacqui

I can't hold my liquor.
Luckily Wahaca
makes margaritas
with very little tequila.

You tell me about Lola:
she is nearly old enough to die.
But we are too young.
May we live to 120!

Whom would we marry
if things got really bad?
It could be any of us hiding in a pigsty,
clinging to a rubber dinghy.

The lilac grows hard by the trash can.
Extinction Rebellion are marching today.
An activist glued herself to a train
in a slightly confusing gesture.

We eat three guacamoles, which is one too many,
and discuss how we would
kill ourselves.
It's easy

in Switzerland, but what if you can't travel?
The exhaust pipe into the car
after getting really drunk, you say,
except if you have an electric car.

Animal Experiments

At the reunion we meet ruin to ruin. Who had children, cancer, or money?

There was a boy who gave up being a thug in order to read. He died in the Australian desert.

Is there a blessing in slowing, forgetting? Chrysanthemums are indestructible, they say.

One boy played piano in the dark. He knew the colours of the pillars at each subway station. Later the colours changed. His knowledge was useless.

Once I saw my own egg on ultrasound. What is the meaning of this?

A girl I knew had read too much and could not write anything down. Everything had already been said.

Work is a lot of work, but not as much as family. It's like people getting off trains, some with skiing equipment.

My mother said *now you are getting As, but later you won't, because you are a girl.* I decided not to be a girl. Recently I am comfortable wearing small earrings.

I find boarding passes in my books like time bombs, my city to your city, your city to my city.

A girl at school carried a brown mouse in her bottle-green sleeves. It developed a skin disease. The vet made her watch its frantic circles after the lethal injection. She has not forgotten the lesson.

I sit on the sofa in my pajamas. The letter says *your test results are normal.*

My mother spoke a local dialect: tough to understand. The yellow chrysanthemum lasted until December. It required very little maintenance.

In the end we all look the same. Only I still recognise myself.

Samosely

You are the last illegal resident
of the exclusion zone.

I speak to you
through fissures

in the concrete shell.
I know you listen.

Radiotropic fungi cover
the reactor walls.

Experts can extrapolate
from silence

back to the explosion.
In twenty thousand years

it will be safe.
You will speak.

Extinction

First
I don't see her
between the trees
fast
rush of stripes
elastic speed.
My gun drops
into the long grass.
I let myself
feel her breath
hot with previous meat.
I accept her
beautiful teeth,
look into that yellow eye
as she feasts.

Light on the Galactic Tide

for Jyothish

You streak through, irregular comet,
surf in from the outer rim of Chamaeleon.
You barely hold on to the center's tenuous gravity,
you are Lemmon, astronomical whim.

You are comical, out of context
in any galaxy. You are brave,
when planets play it safe.
You chisel your singular trajectory into the sky.

Ephemeral guest, you travelled light years
to get here. But you can't stay.
Your visa says *temporary at best.*
Foreigner is the fate that fits.

You are the one cone in an elliptic world,
an acorn in the company of melons.
You ride your eccentric orbit close to the sun.
You know you can never go home.

Irrlicht

We inhabit molasses. Our childhood
house sinks into its basin, shingled in schist.
Mother fells the birches herself.

The terrible truth sits among us in a wheelchair
and knows. Cousin's child is born without fingers.
Mother approves.

The usual will-o'-the-wisp waves me
to the windows of others.
I rise from the fireside and go out

of my mind into the singular twilight.
Fledglings drown in their nests this spring.
Mother makes soup.

A miniature brain grows in a lab in the city.
Immature, what does it think? The sewers are blocked
by fat the size of a factory. Mother makes soap.

Social Isolation Rearing

Sometimes motherless monkey mothers recover from the effects of their motherlessness, if they have offspring. Sometimes the baby monkey itself trains the monkey mother in the basics of monkey affection (babies are efficient bonding machines, even in hopeless cases), and sometimes it doesn't, in which case the motherless mother bites her child's arm off, or worse.

Dump

I don't accumulate,
chuck everything
out. You were a pack rat,
kept your Tesco card
just in case.

At the dump they weigh the car
on the way in
and out,
charge for the difference, the weight
left behind.

I remember—we argued
about holding on to papers,
training manuals for defunct operating systems.

I carry your desk chair across.
Someone unscrews
its only wheel.
Aluminium is the most
precious scrap.

The pillow where the cat slept
goes into a container
connected by conveyor belt
directly to the furnace.
The sign says *don't jump in.*

iv.

does the butterfly remember what the caterpillar knows?

Do It by Leaping

off a mountain, one with a cliff.

From where you live,
take the train (change once) and then a bus to the meadows.

Start early in the day to reach the trailhead before eight.

The cows are out and it's a weekday, so there is silence enough.

If there is another walker share the map you bring,
but don't talk about too many things with him.

Farms   ferns   waterfalls

The mist lifts and you see the end of the trees and the place
up above, its sheer face.

By this time

after hours of walking and grime you crave the water in the flask,
warm from the sun, delicious
as it goes down.

Rosenlaui

We will wash off the mountain
like gravel,
grit.

Torrents will twist
where granite
transitions
to schist. Centuries

later, when glaciers
reverse
their retreat,
we will take
each other's measure.

Moraines will slide across
our human tracks. Casual
ice will crush
slate, scree,
rock.

We will share a terrible and complicated
pleasure.

Slaked,
we won't remain. Mud
will cover us.

The Bad Gardener

smokes on his balcony—
rusted fold-up bike, yellow basil,
withered mint—surveys his scene,
looks down upon the line of refugees
at the Norwegian border,
someone beheading someone else,
a polar bear skeletal on a lonely floe.
I can't keep track of everything.
It isn't fair to blame him
for the beer cans in the muck,
the shootings, unraked leaves
choking the grass. He lights
another one, inhales.

Earth Wakes Up from her Fever Dream

Over salad platters
in Spitalfields you say
this could be the end
(Trump, ISIS, Kim).

I think the million dead
won't make a dent. Remember
the plague wiped out thirty percent,
the human species barely blinked.

A virus might finish us, our own
genetic interventions. We fall
silent, imagine the valley
filled to the brim again, ice

pulsing down to what's left
of the town, in the forest
the next apex predator
giving it a shot.

Humanity

being generally a disappointment—even Euripides
just spills out the lies of power, the slaughter of the other tribe—
where do I turn?

It's not just humans who are selfish,
shortsighted; all of us animals strung up
by our desperate need to live.

The wasp larva eating through the paralysed cockroach,
leaving vital organs to the end; caterpillars in their silk tents
chewing the lime tree down to a skeleton.

The beech sapling only grows in the dapple of light
left behind by the old one falling. But a pumpkin—isn't it a kind of genius,
clicking carbon into place, making sugar of the sun? Isn't it?

Sometimes Someone

steps on a frog
to feel her pop

under his boot
for fun.

Most of us
don't. Most of us

kill millions
because of the inconvenience of not killing.

Trees don't feel pain.
It would be pointless—

they can't jump away,
as a frog sometimes can.

A tree feels no fear,
a tree feels

compassion for the chainsaw,
for the one holding it.

Animal Experiments

I return
from a long journey,
meet a stranger
called Donat
in an empty café
in an empty street.
He tells me
about his delusions,
I tell him
about my grief.
There is nowhere
we have to be.

Fall Fashion

for Horst

I am wearing all my colours
yellow hoodie hand-me-down from Ann
wooly hat with purple pompom
shapeless goretex jacket
(which you hate) in brillant red
and you're a parrot too
next to me in green and blue
around us mouldy leaves and dirty snow
the dark world drawing early night
around her shoulders
grey and tight.

Presence

With precise
eminence
she contemplates
the possibility
of fish
in the churn
of the Limmat
downstream
from the weir
where the late sun
directs its beam
until something
in her brain
motions her to fly
a lighter biomass
than I.

Still Missing

The firs
wear jewels
this morning,

each twig lit
one by
one.

Rime spikes
windward.
I listen.

Fog lifts
off the river
in drifts.

A single finch
mistakenly
sings.

Acknowledgements

My first thanks are to Jacqui Saphra, who is a midwife to this book, as she was to the last. Thank you for our conversations in cafes, the edits and the hugs.

A big thank you to Amy and Jake; I am proud to be a Bad Betty.

Thank you to the editors of *The Poetry Review*, *Under the Radar*, *The Manhattan Review*, *Smiths Knoll*, *The Moth*, *Cimarron Review*, *Magma*, *Beautiful Dragons Press*, *The Stand* and *Flipped Eye*, who previously published some of the poems in this book.

Thank you also to Nii Parkes, Ruth Wiggins, Cheryl Moskowitz, Liane Strauss, Len Joy, Zia Haider Rahman, Carole Allamand, Helen Klein Ross, Maggie Dietz, Stephen O'Connor, Helen Benedict, Philip Fried, Robert Pinsky, Colette Bryce, and especially to John Glenday, readers, friends and mentors.

Thank you to Russ and EunJoo, Jyothish and Purvi, Nawaaz, Vic, Gilda and Hubert Kalf, who are family, and most of all to Horst.

Other titles by Bad Betty Press

2020

poems for my FBI agent
Charlotte Geater

War Dove
Troy Cabida

A Terrible Thing
Gita Ralleigh

Sylvanian Family
Summer Young

bloodthirsty for marriage
Susannah Dickey

At the Speed of Dark
Gabriel Akamo

Rheuma
William Gee

2019

While I Yet Live
Gboyega Odubanjo

She Too Is a Sailor
Antonia Jade King

*And They Are Covered
in Gold Light*
Amy Acre

The Body You're In
Phoebe Wagner

Raft
Anne Gill

Blank
Jake Wild Hall

Alter Egos
Edited by Amy Acre
and Jake Wild Hall

No Weakeners
Tim Wells

Lightning Source UK Ltd.
Milton Keynes UK
UKHW010745191120
373678UK00003B/82